Starting with... Role play

HOPSCOTCH EDUCATIONAL PUBLISHING

Water

Diana Bentley
Maggie Hutchings
Dee Reid

Diana Bentley is an educational consultant for primary literacy and has written extensively for both teachers and children. She worked for many years in the Centre for the Teaching of Reading at Reading University and then became a Senior Lecturer in Primary English at Oxford Brookes University. Throughout her professional life she has continued to work in schools and teach children aged from 5 to 11 years.

Maggie Hutchings has considerable experience teaching KS1 and Early Years. She is a Leading Teacher for literacy in The Foundation Stage and is a Foundation Stage and Art coordinator. Maggie is passionate about the importance of learning through play and that learning should be an exciting and fun experience for young children. Her school's art work has been exhibited in The National Gallery, London.

Dee Reid is a former teacher who has been an independent consultant in primary literacy for over 20 years in many local authorities. She is consultant to 'Catch Up' – a special needs literacy intervention programme used in over 4,000 schools in the UK. She is Series Consultant to 'Storyworlds' (Heinemann) and her recent publications include 'Think About It' (Nelson Thornes) and Literacy World (Heinemann).

Other titles in the series:

Colour and light
Under the ground
Emergency 999
Into space
At the shops
Fairytales
At the hospital
All creatures great and small
On the farm
Water
Ourselves

Other Foundation titles:

Starting with stories and poems:

Self esteem
Self care
A sense of community
Making relationships
Behaviour and self control

A collection of stories and poems

Starting with our bodies and movement

Starting with sounds and letters

The authors would like to thank Jane Whitwell for all her hard work in compiling the resources and poems for the series.

Published by
Hopscotch Educational Publishing Ltd, Unit 2, The Old Brushworks, 56 Pickwick Road, Corsham, Wiltshire, SN13 9BX
Tel: 01249 701701

Written by Diana Bentley, Maggie Hutchings and Dee Reid
Series design by Blade Communications
Cover illustration by Sami Sweeten
Illustrated by Debbie Clark
Printed by Colorman (Ireland) Ltd

ISBN 1 905390 21 1

The authors and publisher would like to thank Chapter One (a specialist children's bookshop) in Wokingham for all their help and support. Email: chapteronebookshop@yahoo.co.uk

Contents

Acknowledgements

The authors and publisher gratefully acknowledge permission to reproduce copyright material in this book.

'There are big waves' by Eleanor Farjeon, from *Blackbird has Spoken*, published by Macmillan. © Eleanor Farjeon. Reproduced by kind permission of the publishers.
'The friendly octopus' by Mike Jubb, from *Sugarcake Bubble Poems*, compiled by Brian Moses, published by Ginn. © Mike Jubb. Reproduced by kind permission of the author. (www.mikejubb.co.uk)
'Seaside' by Shirley Hughes, from *Out and About* by Shirley Hughes, published by Walker Books. © 1988 Shirley Hughes. Reproduced by permission of Walker Books Ltd, London SE11 5HJ.

'Water in bottles, water in pans' by Rodney Bennett and Clive Sansom. © Rodney Bennett and Clive Sansom. Published by kind permission of A&C Black.

Every effort has been made to trace the owners of copyright of material in this book and the publisher apologises for any inadvertent omissions. Any persons claiming copyright for any material should contact the publisher who will be happy to pay the permission fees agreed between them and who will amend the information in this book on any subsequent reprint.

Introduction

There are 12 books in the series **Starting with role play** offering a complete curriculum for the Early Years.

Ourselves	*At the garage/At the airport*
Into space	*Emergency 999*
At the shops	*All creatures great and small*
Colour and light	*Under the ground*
At the hospital	*Fairytales*
On the farm	*Water*

While each topic is presented as a six-week unit of work, it can easily be adapted to run for fewer weeks if necessary. The themes have been carefully selected to appeal to boys and girls and to a range of cultural groups.

Each unit addresses all six areas of learning outlined in the *Curriculum Guidance for the Foundation Stage* and the specific Early Learning Goal is identified for each activity and indicated by this symbol.

Generally, differentiation is achieved by outcome, although for some of the Communication, Language and Literacy strands and Mathematical Development strands, extension activities are suggested for older or more confident learners.

Suggested teaching sequence for each unit

Each week has been organised into a suggested teaching sequence. However, each activity in an area of learning links to other activities and there will be overlap as groups engage with the tasks.

The Core Curriculum: Literacy and Mathematics

Every school will have its own programmes for literacy and mathematics and it is not intended that the activities in the units in this book should replace these. Rather, the activities suggested aim to support any programme, to help to consolidate the learning and to demonstrate how the learning can be used in practical situations.

The importance of role play

'Children try out their most recent learning, skills and competences when they play. They seem to celebrate what they know.'

Tina Bruce (2001) Learning Through Play: Babies, Toddlers and the Foundation Years. London: Hodder & Stoughton.

Early Years practitioners are aware of the importance of play as a vehicle for learning. When this play is carefully structured and managed then the learning opportunities are greatly increased. Adult participation can be the catalyst for children's imaginations and creativity.

Six weeks allows for a role play area to be created, developed and expanded and is the optimum time for inspiring children and holding their interest. It is important not to be too prescriptive in the role play area. Teachers should allow for children's ideas and interests to evolve and allow time for the children to explore and absorb information. Sometimes, the children will take the topic off at a tangent or go into much greater depth than expected or even imagined.

Organising the classroom

The role play area could be created by partitioning off a corner of the classroom with ceiling drapes, an old-style clothes-horse, chairs, boxes, large-scale construction blocks (for example, 'Quadro') or even an open-fronted beach tent/shelter. Alternatively, the whole classroom could be dedicated to the role play theme.

Involving parents and carers

Encourage the children to talk about the topic and what they are learning with their parents or carers at home. With adult help and supervision, they can explore the internet and search for pictures in magazines and books. This enriches the learning taking place in the classroom.

Outside activities

The outdoor classroom should be an extension of the indoor classroom and it should support and enhance the activities offered inside. Boys, in particular, often feel less restricted in outdoor play. They may use language more purposefully and may even engage more willingly in reading and writing activities. In the outdoor area things can be done on a much bigger, bolder and noisier scale and this may connect with boys' preferred learning styles.

Observation in Salford schools and settings noted that boys access books much more readily when there is a book area outdoors.

Resources

Role play areas can be more convincing reconstructions when they are stocked with authentic items. Car boot sales, jumble sales and charity shops are good sources of artefacts. It is a good idea to inform parents and carers of topics well in advance so they can be looking out for objects and materials that will enhance the role play area.

Reading

Every week there should be a range of opportunities for children to participate in reading experiences. These should include:

Shared reading

The practitioner should read aloud to the children from Big Books, modelling the reading process; for example, demonstrating that print is read from left to right. Shared reading also enables the practitioner to draw attention to high frequency words, the spelling patterns of different words and punctuation. Where appropriate, the practitioner should cover words and ask the children to guess which word would make sense in the context. This could also link with phonic work where the children could predict the word based on seeing the initial phoneme. Multiple readings of the same text enable them to become familiar with story language and tune in to the way written language is organised into sentences.

Independent reading

As children become more confident with the written word they should be encouraged to recognise high frequency words. Practitioners should draw attention to these words during shared reading and shared writing. Children should have the opportunity to read these words in context and to play word matching and word recognition games. Encourage the children to use their ability to hear the sounds at various points in words and to use their knowledge of those phonemes to decode simple words.

Writing

Shared writing

Writing opportunities should include teacher demonstration, teacher scribing, copy writing and independent writing. (Suggestions for incorporating shared writing are given each week.)

Emergent writing

The role play area should provide ample opportunities for children to write purposefully, linking their writing with the task in hand. These meaningful writing opportunities help children to understand more about the writing process and to seek to communicate in writing. Children's emergent writing should occur alongside their increasing awareness of the 'correct' form of spellings. In the example below, the child is beginning to have an understanding of letter shapes as well as the need to write from left to right.

Assessment

When children are actively engaged in the role play area this offers ample opportunities for practitioners to undertake observational assessments. By participating in the role play area the practitioner can take time to talk in role to the children about their work and assess their performance. The assessment grid on page 40 enables practitioners to record progress through the appropriate Stepping Stone or Early Learning Goal.

DfES publications

The following publications will be useful:

Progression in Phonics (DfES 0178/2000)
Developing Early Writing (DfES 0055/2001)
Playing with Sounds (DfES 0280/2004)

Water planning chart

Water	Role play area	Personal, Social and Emotional Development	Communication, Language and Literacy	Knowledge and Understanding of the World	Mathematical Development	Creative Development	Physical Development
Week 1	Make an underwater scene	*Respond to significant experiences* Talking about loneliness	*Write labels and captions and form sentences* Making labels for role play area Writing message in a bottle	*Observe and find out about the natural world* Properties of water. Water cycle	*Use mathematical ideas to solve problems* Exploring capacity Experimenting with water levels	*Explore colour and texture in two or three dimensions* Creating background Making fish	*Travel around, under, over, through* Water races Pirate game
Week 2	Add sharks, dolphins and underwater equipment	*Develop awareness of own needs* Talking about bravery and dangers of the sea The 'Titanic'	*Write for different purposes* Listening and discussing story Making posters	*Investigate objects using all senses* Freezing and melting Water experiments	*Use ideas to solve practical problems* Measuring water depth Counting in 4s	*Explore colour, shape and texture in three dimensions* Making sharks and dolphins Making junk models	*Move with confidence, imagination and in safety* Moving to sea music. Playing target game and bucket and coins game
Week 3	Act story of the Rainbow Fish	*Take turns and share fairly* Reading story of Rainbow Fish. Talking about sharing	*Enjoy listening to and using spoken language* Asking children to retell story of the Rainbow Fish Creating new Rainbow Fish story	*Identify features of living things* Reading an information book and discussing it Looking at fresh fish	*Say and use number names* Counting scales Sequencing patterns	*Use imagination in art and design* Making observational drawings Making costumes	*Move with control and coordination* Playing Shipwreck and Flip the kipper games
Week 4	Convert area into beach scene	*Work as part of group* Discussing tasks for creating background scene and the need to take turns	*Write for different purposes* Creating poem with rhythm Exploring the phoneme 's'	*Find out about features of the natural world* Looking at and discussing the coastline and seaside Looking at puddles	*Count reliably up to ten* Counting shells Making groups of 5 Using a number square	*Sing simple songs from memory* Singing songs Making backdrop for role play area	*Move with control and coordination* Playing rock pool games Building sandcastles
Week 5	(Science based week) Add features to beach scene	*Work as part of group, taking turns* Talking about scientists and the need to take turns with experiments	*Sustain attentive listening, responding by asking questions* Devising questions about water properties Writing simple sentences	*Investigate objects and materials* Setting up experiments and observing children's actions	*Use everyday words to describe position* Placing objects in different positions Floating and sinking	*Use imagination in art and design* Making beach scene features.	*Show awareness of space of themselves and of others* Performing action songs Playing a shipwreck game
Week 6	Tropical island beach party	*Form good relationships with adults and peers* Sharing ideas about what they like about the seaside	*Speak clearly and audibly, show awareness of listener* Reciting poem and song to party guests Writing invitations	*Select tools and materials to shape join materials* Preparing food for a party Making 'Undersea jelly', sandwiches and cocktail drink	*Relate addition to combining two groups* Adding two numbers Taking away a number from larger number	*Respond to what they see, hear, smell and touch* Making flower garlands Creating party invitations Setting up outside role play area	*Move with control and coordination* Playing beach games

Water

During this six-week unit, the children will investigate water and the water cycle and consider life under and on the water. They will discuss safety near water and in Week 5 they will become scientists, exploring the nature of water. The role play area will become an undersea scene and, later in the unit, the seashore.

The unit will round off with a 'beach party' and the children will have written invitations for parents, governors and friends of the school.

At the end of the unit, the children will have made:

- a rowing boat
- seaweed, rocks and jellyfish
- sharks and dolphins
- a rainbow fish
- a backdrop for the seaside scene
- a rock pool, harbour and lighthouse
- a submarine

If possible, try to arrange a visit to an aquarium during the unit, or establish a fish tank for the children to observe the movements of fish.

WEEK 1

Starting the role play area

This week the role play area will become an undersea scene. The children will investigate life under the water. They will make the **undersea background**, **eels**, **rocks**, **seaweed** and a variety of **fish**. They will also make the **underside of a boat** to be suspended from the ceiling of the role play area. They will learn about the water cycle and discover why water is important. Adults should participate in the role play area by taking on the role of a diver exploring the ocean floor. They will write a message and put it in a bottle. NB: place a **tape recorder/CD player** in the role play area.

The sea

Resources

Photocopiables:

Poems and songs (page 32)
Making the role play area background (page 33)
Instructions for making fish (pages 33–34)

Fiction books:

Mr Gumpy's Outing by John Burningham, Red Fox (0 099408 79 1)
Old Shell, New Shell by Helen Ward, Templar Publishing (1 840119 03 9)
Little Beaver and the Echo by Amy MacDonald, Walker Books (0 744523 15 X)

Non-fiction books:

Under the sea by Alastair Smith, 'Flap Books' series, Usborne (0 746055 82 X)
A Drop in the Ocean: The Story of Water by Jacqui Bailey, 'Science Works' series, A&C Black (0 713662 56 5)
Oceans and Seas by Nicola Davies, 'Kingfisher Young Knowledge ' series, Kingfisher Books (0 753409 45 3)
Little Nippers: Let's Get Moving Under the Sea by Victoria Parker, Heinemann Library (0 431164 84 3)

Poetry:

Commotion in the Ocean by Giles Andreae, Orchard Books (1 841211 01 X)

Music and songs:

Tom Thumb's Musical Maths by Helen MacGregor, A&C Black (0 713649 71 2)
Michael Finnigin, Tap your Chinigin by Sue Nicholls, A&C Black (0 713647 16 7)
The Wheels on the Bus, CYP, CD (1 857817 15 X)
CD of ocean music, for example, *Peaceful Ocean Surf,* SPJ Music (B000021YLC) or *Voyage on the Ocean: Sounds of Nature,* Intercontinental (B000021YE7)

Materials:

- Pictures and posters of under the sea and undersea animal life
- Sand coloured fabric, other fabrics
- Green garden netting
- Simple posters or diagrams to illustrate the water cycle
- Kettle
- Tissue paper, coloured and plain Cellophane
- Sponge
- Kitchen paper, greaseproof paper, tinfoil
- Bubble wrap
- Bottles with corks
- Tape recorder
- Jugs, sieves
- Globe

Personal, Social and Emotional Development

Respond to significant experiences, showing a range of feelings when appropriate.

Robinson Crusoe

- ❑ Tell the children the beginning of the story of Robinson Crusoe, who was shipwrecked and ended up on a desert island. Talk about how lonely he must have felt. What could he find to eat? Challenge the children to share ideas about how he could escape from the island. Ask them to imagine being stranded on a desert island; what one thing would they like to be able to take with them?

Circle time

- ❑ Remind the children of how lonely Robinson Crusoe felt on the desert island. Read *Little Beaver and the Echo* (see Resources) or a similar story about being lonely. Go round the circle, asking children what they do when they feel lonely: 'When I feel lonely I …' Ask them if they have ever felt lonely at school. Discuss what can be done to ensure no-one is lonely at playtime.

Knowledge and Understanding of the World

Observe, find out about and identify features in the natural world.

Water

- ❑ Tell the children that in this unit they are going to be studying water and life in the sea. In small groups, give each child a clear, plastic cup of drinking water. Ask them: What does the water looks like? (Clear; can see through it.) What does it smell like? What does it taste like? Does some water taste different? (Sea water; bottled fizzy water.) Tell them to gently move their cups. Ask them what happens to the water. Then ask them to gently blow the water. What can they see? Tell them to pour their cup of water into a basin. What can they hear as the water pours?

- Explain that water is a liquid. What other liquids can they think of? Discuss what we use water for. Explain that about half our weight is the weight of the water in our body. Ask them why water is so important. People and animals and plants need water to live. Without it they would die. We need water to keep us healthy. We should drink plenty of water each day.
- Explain that some parts of the world do not have enough water, so rivers and lakes dry up, plants do not grow and animals die. Ask the children where they think water comes from. Look at a globe. Talk about the area of the world that is sea.

Water cycle

- Explain that clouds are full of tiny drops of water that fall as rain to the ground. The water soaks into the ground or falls into rivers which flow down to the sea. When the rain stops and the sun shines, the heat from the sun makes the water evaporate into mist, which rises into the air and makes new clouds, which will fall as rain again. This is called the water cycle. Draw a simple version of the water cycle.

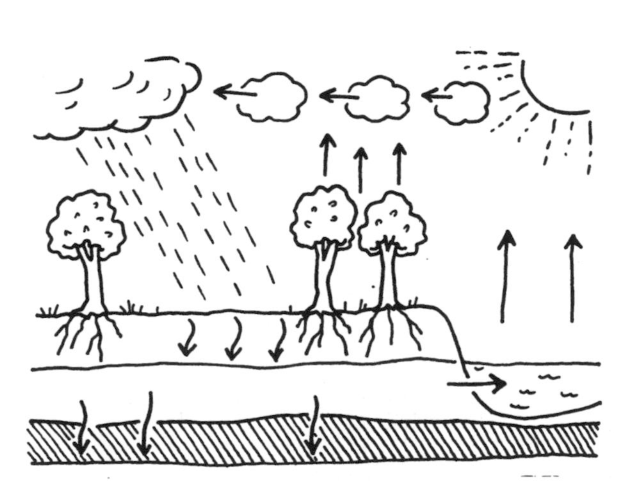

Water experiments

- Explain evaporation and condensation in simple terms; for example, show how water heated in a kettle makes steam or a mist and that when it cools it becomes droplets of water again. Hold a cold spoon near the steam.
- Linking with the activity in Mathematical Development (droplets of water on finger tips), try letting droplets fall onto different surfaces, such as kitchen towel, tissue paper, sponge, greaseproof paper, foil and fabrics. Tell the children to observe what happens. Do the droplets change? Why do some disappear? Talk about absorbency of materials. Which would be best for soaking up or mopping up spilt liquids? Try them!

Coral reefs

- Talk to the children (and look at pictures) about coral reefs. Explain that they will make coral to add to the decoration of the role play area.

Creative Development

Explore colour, texture, shape, form and space in two or three dimensions.

Making the role play background

- See page 33 for suggestions for making the background for the undersea role play area.
- See pages 33 and 34 for suggestions for making fish to decorate the role play area.

Music and song

- Play a CD of ocean sounds (see Resources).

The sea

Mathematical Development

Use developing mathematical ideas and methods to solve practical problems.

Exploring capacity

- Place different sized containers in the water tray. Include plastic bottles, narrow necked bottles, jugs, funnels and sieves. Encourage the children to use early mathematical language related to capacity, such as 'full', 'empty' and 'half full'. Ask them to estimate, for example, the number of small full containers needed to fill a larger one.

Extension

- Provide different shaped containers and ask the children to estimate how many small full containers will be needed to fill them.

Properties of water

- Ask the children to pour water into a bottle. What happens? What could you use to make it easier? Experiment with pouring water from varying heights. What happens? Did all the water go into the container? Can you fill a sieve with water? Why not? Look at the pattern as the water flows through the holes in the sieve. Try holding a clear container of water at different angles. Does the level of the water stay the same? Dip your finger into the water. Can you make drops hang from your fingers? What shape are the drops? Can you draw droplet shapes?

Counting

- Ask the children to count tentacles on the octopus. What other animal has eight legs?
- Read the poem 'The friendly octopus' (see page 32) and count with the children.

Shapes

- Sing 'A pirate went to sea, sea, sea' from *Tom Thumb's Musical Maths* (see Resources) and discuss the shapes the pirate sees – circles, triangles and squares.

Communication, Language and Literacy

Write labels and captions and begin to form simple sentences, sometimes using punctuation.

Writing labels

- Make the label 'Under the sea' for the whole role play area. Draw round large letter templates and ask the children to decorate each letter with sea creatures.
- Talk to the children about labels for the role play area, naming 'fish', 'octopus', 'rocks' and 'coral'. Say each word clearly and ask them to identify the initial sound. Write this letter and then help them to identify the next sound they can hear. Work through each word until it has been spelled. Take the labels into the role play area and discuss with the children where they should be displayed.

Pictures and labels

- Give each child a cloud shaped piece of paper and ask them to draw pictures of things related to the water cycle, such as clouds, rain, river and sea. Write labels for the pictures using highlighter pen and let the children go over it with pencil.

Extension

- Give pairs of children a larger cloud shaped piece of paper. Ask them to draw and write about the things they have found out about water on small pieces of paper and stick them onto the cloud shape. Talk about this first and help them to formulate their ideas into sentences and then share some ideas on the class whiteboard. Scribe some of the tricky words and leave for the children to refer to as they work.

Message in a bottle

- Before working with the children, write a message, place it inside a bottle and seal it with a cork. (This could be something like: Help! I am John. I am stranded on an island. Please rescue me!) Show the children how the bottle floats on water in the water tray. Lift the bottle out and 'discover' the message inside. Explain that it is from a man who has survived a shipwreck. Read the message.

Tell the children to imagine that they are stranded on a desert island and they are going to send a message in a bottle. Talk over some of the children's ideas and demonstrate writing messages on the class whiteboard. Then give each child a bottle shaped piece of paper with an area for writing. Encourage them to write their own messages. (Give support according to ability.) Display the bottles in or around the role play area. Ask the children to read their messages into a tape recorder. Leave the tape in the role play area.

Desert islands

- Draw on a large sheet of paper an outline of a simple map of an imaginary desert island. Talk about the various features. For example, palm trees, caves, beach, forest, hill, river and coral reef, and draw and label these on the map. Display the map in the role play area and encourage the children to act out being shipwrecked and landing on a desert island, using the vocabulary on the map.

Music and song

- Sing the nursery rhyme 'One, two, three, four, five; once I caught a fish alive.'
- Sing 'The day I went to sea' from *The Wheels on the Bus* (see Resources).
- Sing 'Thunderstorm' from *Michael Finnigin, Tap your Chinigin* (see Resources).

Physical Development

Travel around, under, over and through balancing and climbing equipment.

Pirates shipwreck game

- Show the children how to set out large apparatus in the hall: mats, benches, skipping ropes and hoops. It should be set out so that it is possible to move around the hall without 'touching' the ground. For example, a bench linked to a mat, linked to another mat by a length of skipping rope, linked to a hoop, linked to a mat and so on. Try to set out the apparatus to present as many different paths or options for movement as possible. The children pretend they are pirates and move across the apparatus, using control, coordination and balance. They should begin from different starting positions and avoid queuing by finding different paths. If space is limited, reduce the number of children working at one time. Others can observe and comment on ways their friends have been able to negotiate the obstacles.

Cutting skills

- See Creative Development.

Movement

- Direct the children to move as sea creatures: moving through all the spaces; darting like small fish; wriggling like an eel; swimming gracefully; as if in a shoal (in small groups); slowly, waving tentacles like an octopus. Accompany the music with some sea sound effects (see Resources).

Outdoor play

- Have a water race. You will need buckets of water, sponges and smaller containers, clear if possible. Divide the class into teams. Each team has a bucket of water and a sponge. Place a smaller container, at a distance, opposite each team. The aim is for children to take it in turns to soak the sponge and race to fill the container. The first team to fill their container wins.

Water

WEEK 2

The role play area

During this week the children will add **sharks, dolphins, oxygen tanks, iceberg pictures** and **junk model submarines** to the role play area. They will investigate the properties of water – liquid and solid. They will consider the power of the sea and its dangers, and learn about the 'Titanic', the passenger liner which hit an iceberg and sank in the Atlantic many years ago.

Resources

Fiction books:

Smiley Shark by Ruth Galloway, Little Tiger Press (1 854308 62 9)
Mr McGee Goes to Sea by Pamela Allen, Puffin (0 140544 03 8)

Non-fiction books:

Little Nippers: Let's Get Moving Under the Sea by Victoria Parker, Heinemann Library (0 431164 84 3)
Under the Sea by Fiona Pratchett, Usborne Publishing (0 746045 43 3)
Survivors – The Night the Titanic Sank, Dorling Kindersley Readers Level 2 (0 751314 73 0)

Music and songs:

Recording of 'Yellow submarine' by the Beatles
'Who's afraid?' from *The Handy Band* by Sue Nicholls, A&C Black (0 713668 97 0)
'Getting dressed' from *Big Steps, Little Steps,* Kindescope, www.kindescope.com

IT:

National Geographic Video – 'Secrets of the Titanic' (ASIN: B00004CJY9)
http://octopus.gma.org/space1/titanic.html

Materials:

- Flippers, snorkels and masks (NB: sterilise the mouthpiece between each child)
- Buckets and spades
- Binoculars
- Magazines, such as old National Geographic, showing pictures of the Antarctic
- Large fizzy drinks bottles, fizzy drinks
- Black tape and lengths of webbing or wide braid
- Pictures and posters showing icebergs, the 'Titanic' and submarines and divers
- Raisins
- Ice cube trays
- Balloons
- Plasticine
- Old pen tops, polythene bags
- White blown vinyl wallpaper
- Dressing up clothes for the passengers on the 'Titanic'
- Tinfoil

Knowledge and Understanding of the World

Investigate objects and materials by using all of their senses as appropriate.

Freezing and melting

- ❑ Ask the children if water is always a liquid. Talk about the different forms of water, such as rain, mist, hail, sleet, snow and ice. Explain that when water is frozen it becomes a solid – ice.

Experiments

- ❑ Fill an ice tray with water. Place it in a freezing compartment and show it to the children over periods of time. How has it changed? Place an ice cube on a plate and observe the melting process from solid back to liquid. Make a note of how long it took for the ice cube to melt. How did it change in shape as it melted?
- ❑ Place an ice cube in the fridge and compare the time it takes to melt with a cube left at room temperature.
- ❑ Fill a container, such as a lidded margarine tub, with water and freeze it. Observe how the water takes up more space when it is frozen – the ice forces the lid. Alternatively, fill a plastic drinks bottle with water and leave, upright, in the freezer (no need to put on cap).
- ❑ Let the children hold an ice cube and describe what it feels like.

Balloon iceberg

- ❑ Fill a balloon with water and freeze it. Remove the balloon from the ice. Place the ice in a clear tank of water. Observe how it floats. Ask the children if they can see how much of the ice is under the water's surface. Relate this to the icebergs floating in the sea. Observe the changes to the ice balloon as it melts. Ask the children if the ice balloon changes its shape as it melts. Does it move in the water?

Submarines

- ❑ Look at pictures of submarines. Explain that submarines have tanks that are filled with sea water to make them dive and that air is pumped into the tanks to push the water out to make the submarine lighter. This helps them to rise towards the surface.

Making a submarine

- ❑ Take a pen top and seal the hole at the top with Plasticine. Stick a blob of Plasticine on the end and test to see if it will float upright in water. Fill a plastic drinks bottle with water and float the submarine. Screw the lid on tightly. Squeeze the bottle firmly. The submarine will sink to the bottom because water is forced into the pen top. Stop squeezing and the submarine will rise to the top again because water leaves the pen top, making it less dense.

Water in a bag

- ❑ Almost fill a polythene bag with water. Let the children hold the bag and gently squeeze it. See how its shape changes. Freeze it. Can you change the shape?

Ice floating

- ❑ Drop a coloured ice cube (use food colouring) into a clear container of hot water. Does the ice cube float or sink? What happens as the ice melts?

Raisins and fizzy liquid

- ❑ Take a jar of fizzy drink (carbonated water or lemonade) and drop in a handful of raisins. Watch and see what happens. Look at the bubbles sticking to the wrinkles on the raisins as they rise and fall in the liquid. When the bubbles on the raisins burst at the surface, the raisins sink.

Personal, Social and Emotional Development

Have developing awareness of their own needs, views and feelings and be sensitive to those of others.

Circle time

- ❑ Talk about bravery and courage. Explain that people are brave when they do something that they are frightened to do or when they put themselves in danger to help other people, such as jumping into water to rescue someone.
- ❑ Pass a hand puppet round the circle and ask the children if they can think of times that they, or a member of their family, have been brave. Relate to water and swimming.
- ❑ Sing 'Who's afraid?' from *The Handy Band* (see Resources).

The Titanic

Discussion

- Talk about the dangers of the sea, how it is very powerful and dangerous. Talk about lifeguards, lifeboats and lifebelts. When we go to the swimming pool, there are lifeguards who watch all the time. Ask the children why they think the lifeguards are there. Explain that they are ready to help people in difficulty. Ask them if they have noticed lifebelts on the side of the pool.

Titanic

- Tell the children about the 'Titanic', a huge passenger liner that hit an iceberg in 1912 and was so badly damaged that it sank in the freezing water of the Atlantic Ocean. Show the class the Atlantic Ocean on a globe and point out where the 'Titanic' was travelling from and to. Tell them that the ship was not carrying enough lifeboats for all the passengers and crew. Women and children were put into the lifeboats first. Do they think that was fair? The rich people in first class were helped before the poor in steerage. Do they think that was fair? How do they think the passengers and crew felt when they knew that the ship was sinking? If possible, show selected excerpts from a video about the 'Titanic'.

Communication, Language and Literacy

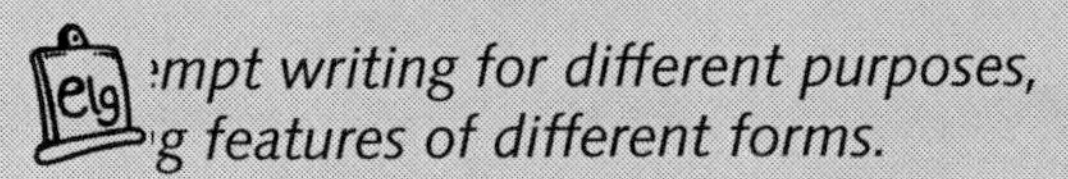

mpt writing for different purposes,
g features of different forms.

Listening

- Read a fiction story about a shark or other sea creature (see Resources). Discuss the story with the class and ask them which part they liked the best.
- Look at a non-fiction book about sharks and dolphins (see Resources). Ask if any of the children have ever swum with a dolphin or know someone who has. Tell them that very few sharks are dangerous to swimmers but lots of stories have been told about man-eating sharks and this has given them a bad name.
- Tell or read the story of the 'Titanic' (see Resources). Discuss with the children the dangers of the icebergs and the conditions of the sea. If possible, look at video footage about icebergs and selected extracts from a film of the 'Titanic'.
- Look at a non-fiction book about icebergs (see Resources) and draw attention to any new vocabulary, such as 'glacier', 'ocean', 'frozen water', 'jagged', 'craggy', 'cold' and 'enormous'.

Posters

- Make a class poster about icebergs. Ask the children for words to describe the icebergs and write these on the board. Talk about the letter formation and letter sounds as you write.
- Give each child some ice floe shapes roughly cut from good quality white paper. Ask them to record words and facts they know or have found out which describe ice and icebergs. Support the children by writing their selected words in yellow highlighter and ask them to trace over the letters or encourage them to copy the words from the board.

Extension

- More able children could write further information about icebergs, such as 'When an iceberg floats in the sea, a lot of it is under the water and you can't see it,' or 'The Titanic was a big ship that sank after it hit an iceberg.' Children could illustrate their posters with drawings or find pictures to cut out of magazines.

Recording

- Support the children as they record their observations, in pictorial form, of the experiments on freezing water and melting ice (see Knowledge and Understanding of the World). Some children could annotate with single words or phrases.

Song

- Sing or play 'Yellow submarine' by the Beatles or 'The big ship sails on the alley alley o'.

Mathematical Development

Use developing mathematical ideas to solve practical problems.

Measuring

- Talk about water depth in a swimming pool: the shallow end; the deep end; being out of one's depth when the water is deeper than you are tall.
- Pour water into the water tray. Can you find out how deep it is? (Use a ruler or stick and mark the depth.) Pour more water into the tray and measure and compare the depth. Pour water into different containers and compare the depths. You may wish to add a little food colouring to the water.

Counting

- Explain that the 'Titanic' had four funnels. Ask the children what other 4s they can think of. For example, four legs on a dog, four fingers next to my thumb and four wheels on a car.

Extension

- Count in 4s, marking them on a number line or number square.

Creative Development

Explore colour, texture, shape, form and space in two or three dimensions.

Making sharks

- Show the children pictures of sharks. Draw a shark on a large piece of card and cut it out. Stick loosely scrunched newspaper over the shape with PVA glue and layers of kitchen paper. Make sure that you cover over the whole shape with at least one layer of kitchen paper to finish. When dry, paint pale grey. Add eyes and teeth with black and white paint. When the paint is dry, paint with slightly diluted PVA glue to give a sheen over the shark's surface. Don't worry if the surface looks cracked or curls as this will give texture. Display the shark across a corner in the role play area.

Making dolphins

- Show the children pictures of dolphins. Ask them to draw a dolphin on card and cut it out. Cover the whole shape with a sheet of kitchen foil and tuck the edges of the foil underneath. Smooth the foil. Add some eyes using a black permanent marker pen. Display them in groups, in and around the role play area.

Making diving equipment: oxygen tanks

- Tape two large, empty fizzy drinks bottles together with tape and cover them with kitchen foil. Using strong duct tape, add braid or webbing straps to fit over the shoulders. Store in a box in the role play area with snorkels and masks.

Submarines

- Show the children pictures of submarines. Let them make their own models from junk, tubes and boxes. Display one in the role play area and the others around the outside.

The Titanic

Iceberg pictures

- Give each child a piece of blue card or paper. Have available a roll of white, blown vinyl wallpaper. Tear or cut a large iceberg shape (or shapes) and glue it to the blue paper. Sponge print the sea with thin dark blue and green paint to show the depth of the iceberg under the water. Mount the pictures around the role play area and the classroom.

Junk models of the Titanic

- Show pictures of the ill-fated ship. Provide junk materials, such as boxes and tubes, and encourage the children to build their own versions of the 'Titanic'. Display them in or around the role play area.
- Display one of the models on the sea bed in the role play area to show how the ship broke in two as it began to sink.

Physical Development

Move with confidence, imagination and in safety.

Movement

- Using sea music (see Resources), practise swimming strokes and pretend to be divers under the sea.
- Using mats, ask the children to lie on their tummies with their hands by their sides. They should imagine they are wearing flippers and move their legs smoothly up and down?
- Try pretend swimming front crawl, breaststroke, perhaps even butterfly and then backstroke.

Dressing skills

- Encourage the children to dress and undress independently after Physical Development. Give them time to practise more difficult skills, such as tying, and encourage them to help each other.

Music and song

- Sing and join in with one of the songs from *Big Steps, Little Steps* (see Resources).

Outside activities

- Bucket and coins game. You will need a bucket of water and some real coins.

  Place a small coin at the bottom of the bucket of water. Ask the children to drop coins into the bucket to try to cover the coin at the bottom. Observe what happens as the coins drop through the water. Why is it difficult to aim the coins? Is it better to drop the coin flat or on its edge? Is it better to drop the coin from a height or near the surface?

Target game

- Aim water pistols at a target which is either drawn in chalk on a wall or, for example, a skittle on the ground. Does the water spurt out in a straight line or does it spurt out in a curved line? Is it easy to hit the target?

Cutting skills

- See Creative Development.

WEEK 3

The role play area

During this week the activities in the role play area are based on the story of Rainbow Fish. The children will add rainbow fish to the role play area and have opportunities through drama (hot seating) and imaginative play to explore sharing and feelings. They will make a **rainbow fish**, **small fish**, **fish hats**, an **octopus costume**, badges for little fish and **portholes**.

Resources

Photocopiables:

Poems and songs (page 32)
Instructions for making fish, octopus, fish costumes, portholes, fish hats (pages 34 and 35)

Fiction books:

The Rainbow Fish by Marcus Pfister, North – South Books (1 558580 09 3)
Rainbow Fish and the Sea Monsters' Cave by Marcus Pfister, North – South Books (0 735815 36 4) (or other Rainbow Fish books)
The Fish Who Could Wish by John Bush, OUP (0 192722 40 9)

Non-fiction books:

Sea Animals by Clare Llewellyn, 'I Know That' series, Franklin Watts (0 749651 69 5)
In the sea by Vic Parker, 'Me and My World' series Franklin Watts (0 749656 53 0)

Music and song:

The Handy Band by Sue Nicholls, A&C Black (0 713668 97 0)

Materials:

- The Rainbow Fish lotto cards, available from educational catalogues
- Rainbow Fish hand puppet (relatively expensive but worth the investment!), Puppets By Post
- Shiny paper (coloured foil)
- Fresh fish from the supermarket
- Chair or stool
- Length of blue or green fabric
- Felt fabric; see animals for colours
- Funnels, sieves
- Colander
- Coffee filter
- Flannel or towelling

The Rainbow Fish

Personal, Social and Emotional Development

Taking turns and sharing fairly.

Circle time

- ❑ Read the story *The Rainbow Fish* (see Resources). Talk about the story. Ask the children if they can describe how Rainbow Fish felt as he swam in the sea. (He felt proud because he looked so beautiful.) Ask: What makes you feel proud? What happened when the little fish asked Rainbow Fish for one of his shiny scales? (He refused.) How did the little fish feel and what did he do? Why do you think the little fish did not want to have anything to do with Rainbow Fish? Rainbow Fish was lonely, so who did he talk to? (The wise old octopus.) What did the wise old octopus say Rainbow Fish should do? (He should share his scales.) How did Rainbow Fish feel then? How did Rainbow Fish feel once he had shared his shiny scales?
- ❑ Talk about sharing. Allow the children to reflect on why it is good to share.
- ❑ Sing 'My turn, your turn' from *The Handy Band* (see Resources).

Drama: hot seating

- ❑ In the role play area, model hot seating by taking on the role of either Rainbow Fish or the wise old octopus. Encourage groups of children to take on the other roles, asking questions of the character in the hot seat. For example, a little fish might ask the wise octopus why he suggested Rainbow Fish should give away his shiny scales.

Mathematical Development

Say and use number names in order, in familiar contexts.

(If you have a Rainbow Fish puppet, use it to give instructions for the following activities or wear a Rainbow Fish 'poncho' – see Creative Development.)

Counting

- ❑ Either use coloured counters or cut out different coloured scale shapes and laminate them. Ask individual children if they can give you; for example, two green scales, six blue scales and so on.

Additing and Subtracting

- ❑ Challenge the children to find three red scales and four blue scales. How many scales do they have altogether? If they have six shiny yellow scales and you borrow two of them, how many will they have left?

Extension

- ❑ Show the children how to record these as written sums; for example, $3 + 4 = 7$ and $6 - 2 = 4$.

Sequencing patterns

- ❑ Make a pattern of scales – for example, red, red, blue, red, red, blue – and ask the children to identify and continue the pattern. Ask some of them to record the pattern on squared paper.

Music and song

- ❑ Counting down from 10. Sing the following song to the tune 'Ten green bottles'.

  Ten shiny scales on the Rainbow Fish
  Ten shiny scales on the Rainbow Fish
  And if one shiny scale was given to a little fish
  There'd be nine shiny scales on the Rainbow Fish
  (And so on.)

Communication, Language and Literacy

Enjoy listening to and using spoken and written language and readily turn to it in their play and learning.

Listening

- Reread the story *The Rainbow Fish* to the class (or another Rainbow Fish story). Invite the children to retell the story on tape. Some of them might prefer to use the pictures in the book to help them with sequencing the events. Place these recordings in the role play area for the children to use in their play. Show them how to record and how to play back the recordings.

Poetry

- Share with the children the poem 'If you ever' (see page 32). Encourage them to learn the poem by heart.

Oral language

- When the children are familiar with the story of the Rainbow Fish, help them to make up a new story about the little fish. Support them with the story's creation by offering possible alternative events and endings. (For example: Does he go into the dark cave or does he swim to ask the whale for help? Does he become friends with the shark or does the shark swim away and never return?) Let the children select and add details to the story.

Extension

- Talk to the children about the stories they have created. Help them to sort out the order of events and let them record these as drawings. Then ask them to write a sentence for each drawing. They might like to do this on fish shaped paper.

Writing labels

- Show the children a fresh fish and discuss with them the vocabulary that describes the different parts – fins, tail, eyes, mouth, scales and gills.
- Explain that you are going to draw a fish and label it. Ask them to identify the different parts on your fish and encourage them to tell you what letter sound the words begin and end with.
- Give each child a fish shaped piece of paper and ask them to draw and label a fish. Display these as a shoal in the role play area.

Knowledge and Understanding of the World

Find out about, and identify, some features of living things.

Reading for information

- Share one or two non-fiction books about fish. Discuss with the children what they have learned that they did not know before. Encourage them to browse through books with pictures of fish in the ocean. Encourage them to talk about the different shapes and colours.

Observing real fish

- Show the children a live fish in a tank. Observe the way it swims. What does it do with its mouth? What does it do with its fins and tail?

Experiments

- Making dirty water clean. Explain that fish in the tank need clean water to survive, just as we need clean water to drink or for our bath. Our water is stored in reservoirs. It is cleaned before it is ready for us to drink. Tell them that they are going to find out how to make dirty water clean.

  You will need: clear containers, such as plastic cups, a selection of muslin/flannel/small piece of plastic (bin liner)/cotton handkerchief, sieve, colander, elastic bands, lidded plastic container, plastic bowl and spoon.

  Make a mix of soil and water in the lidded plastic container (include a few small stones). Shake well.

The Rainbow Fish

Experiment with pouring the mixture through the colander over the bowl. What happens? Try pouring the same mixture through a sieve. What happens? Which cleaned the water better?

Cover the tops of clear containers with different materials and secure with elastic bands. Spoon soil and water mix onto the different materials. Which materials clean the water best? What happened when you spooned the dirty water onto the plastic? Which materials allowed water to go through them? Which material was waterproof?

Looking at and touching fish

- Try to obtain a selection of fresh fish from the supermarket or fishmonger. For example, small whole fish such as sardine, whitebait and squid if you can get them. Ask the children to look closely at the fish and touch them. NB: be aware of allergies! Can they feel the scales? Ask them to run their fingers from head to tail, then from tail to head. Which way feels rough? What colours can you see on the scales? Help the children to describe the fins; for example, 'They look transparent,', 'They have lines in them,' and 'They are made of bony material.'

Creative Development

elg *Use their imagination in art and design.*

Drawing and painting

- Encourage the children to make observational drawings of fresh fish (see Knowledge and Understanding of the World).

Rainbow Fish

- Ask a child to draw a large Rainbow Fish shape on blue card. Invite others to decorate it with shiny scales, leaving some gaps where he has given scales away.
- Make fish, octopus and rainbow fish costumes, portholes and fish hats (see pages 34–35).

Physical Development

Move with control and coordination.

Movement games (indoors or outside)

- Shipwreck game – make signs 'North', 'South', 'East' and 'West' and display these on the walls in the hall or in the playground. For younger children, just have 'north' and 'south'. Explain and demonstrate the following instructions and actions (use a few only to start with): for example, 'North' – run towards the 'north' wall; 'Captain's coming' – stand and salute; 'Man overboard' – hold nose and jump forward into the water; 'Wave coming over' – lie down flat on tummy; 'Climb the rigging' – pretend to climb up hand over hand; 'Man the lifeboats' – sit on floor and row; 'Up aloft' – lie on back with legs in the air; 'Freeze!' – stay absolutely still.

  The object of the game is to follow the instructions as quickly as possible. When the children are familiar with the game, speed up the instructions and try to catch them out.
- Flip the kipper game – cut four fish shapes out of thin paper. Have either four pieces of card or four small trays for flapping. Mark start and finish lines. The children have to flap the card or trays to make the fish move. The first to go over the finish line is the winner.

Fine motor skills

- See Creative Development – cutting skills.

WEEK 4

The role play area

During this week the role play area will be converted into a beach scene. The children will begin to investigate the coastline, waves, tides and life at the seaside. They will make a large **backdrop** for the role play area and make **sandcastles, rock pools, crabs, starfish**, a **breakwater** and a **harbour**.

Resources

Photocopiables:

Poems and songs (page 32)
Instructions for making seaside backdrop, sandcastles, rock pool and so on (pages 35 and 36)

Fiction books:

Lucy and Tom at the Seaside by Shirley Hughes, Puffin (0 140544 59 3)
Sharing a Shell by Julia Donaldson, Macmillan (1 405020 47 4)
Brilliant Boats by Tony Mitton, 'Amazing Machines' series, Kingfisher (0 753407 21 3)

Non-fiction books:

Seashore by Lynn Huggins-Cooper, 'First-hand Science' series, Franklin Watts (0 749648 61 9)
Seashore by R and L Spitsburgh, 'Wild Britain' series, Heinemann Library (0 431039 03 8)
At the Seaside by Tony Pickford, 'What is it Like Now?' series, Heinemann Library (0 431150 09 5)
High Tide, Low Tide by Mick Manning, Franklin Watts (0 749641 81 9)

Music and song:

Seaside Retreat, CD, Solitudes (ASIN B00006L3PX)
'I do like to be beside the seaside', www.allthelyrics.com

Materials:

- 'Aquaplay' (www.toysdirecttoyourdoor.co.uk) or other clip together series of canals, locks etc
- Boat shaped water tray boat or large box
- Set of oars (or make with poles and card)
- Buckets and spades
- Fishing nets
- Sun hats, sunglasses
- Shells
- Pictures and posters of the sea and seashore, rock pools and sea creatures

The beach

Personal, Social and Emotional Development

Work as part of a group or class, taking turns and sharing fairly.

Circle time

- ❑ Tell the children about a day you spent at the seaside (imaginary or true). Describe how you got there and what you saw and did. Ask the children if they have ever been to the seaside. How did they get there? Who did they go with? What did they do there? What did they take with them to play with? Was the sea rough or calm? What colour was it? What did they see on the beach?

Discussion

- ❑ Talk to the children about what makes us happy. Tell them to turn to a partner tell them of something they have done which made them really happy. It might have been going somewhere special or meeting someone special. When they are ready, ask them to tell the group their special happy memory. Make a list of all the things they mention and display this near the door. As children leave the classroom, choose one of the 'happy memories' and encourage everyone to think happy thoughts as they move on.

Working together and taking turns

- ❑ Talk to the children about how the backdrop to the role play area will need to be made with everyone cooperating. Some will make footprints in the sand; others will help with the painting of the sky and sea. Together, their work will create the seaside scene.

Music and song

- ❑ Sing 'The sun has got his hat on' (from CD of popular children's songs).

Mathematical Development

Count reliably up to ten everyday objects.

Counting

- ❑ Use shells as counters on a number line.
- ❑ Make groups of five shells in as many different patterns as possible.
- ❑ Make 20 starfish out of orange card, small enough to fit on numbers on a 100 square. Laminate them. Put a blob of sticky tack on the back of each starfish. Count in fives and place a starfish on every fifth square.

Extension

- ❑ Start on, for example, 2 and count along in 5s. Can they see a pattern?

Sorting

- ❑ Sort shells by colour, texture or type.

Additing and subtracting

- ❑ Ask the children to count how many starfish they have in the role play area. If they took away two fish how many would be left? If they took away another one how many would be left?

Counting rhyme

- ❑ Sing, to the tune of 'Ten green bottles':

  Six sea shells on the seashore (repeat)
  And if one sea shell should simply float away
  There'd be five sea shells on the seashore.

Communication, Language and Literacy

Attempt writing for different purposes, using features of different forms (poetry).

Listening

- ❑ Select one of the poems from page 32 (for example, 'The seaside' by Jo Peters) and read it to the children. Ask them to help you and encourage them to join in with the verses. If possible, do action movements to accompany the poem.
- ❑ Read a story about the sea; for example, *Lucy and Tom at the Seaside* (see Resources). Discuss what happens in the story and ask the children to help you to retell the story in sequence.
- ❑ Listen to some music of the sea (see Resources). Ask the children to describe how the music makes them feel. Think of some words to describe the sounds you can hear – 'splashing', 'lapping', 'crashing', 'spraying' and so on. Write the words on the board, talking about the writing process as you write.

Writing – poetry

- ❑ Using the words from above, talk to the children about making up a poem. Stress the need for rhythm and not rhyme; for example: 'Crashing and rushing, creeping and flowing, lapping and spraying … are the waves on the beach.' Write the poem on the board and discuss the rhythm. Does it sound a little like the waves? Ask the children to each draw a picture of waves and then help them to write one line of their poem under their picture. Make this into a class book or display on the classroom wall as a poem.

Extension

- ❑ Help the group to make up a poem orally. Explain how a similar line at the beginning or end of each verse helps to give the poem form. Invite one or two children to write out the poem and then ask the whole group to record a reading of the poem on tape. Play this to the rest of the class and leave it in the role play area for use on other occasions.

IT

- ❑ Invite the children to record their memories of a day at the seaside.

Speaking

- ❑ Teach the class the famous tongue twister: 'She sells sea shells on the seashore; the shells that she sells are sea shells I'm sure.'

Letter sound 's'

- ❑ Tell the children that they are going to say a rhyme and add as many children's names beginning with 's' as they can: Sally saw sea shells on the seashore. Along came a big wave, and then there were there no more. Sam saw sea shells on the seashore. Along came a… Continue with other names; for example, Susan, Salina, Salim, Suzy, Stephen, Salma, Sultan, Stuart, Sarah, Surinder.

Knowledge and Understanding of the World

Observe, find out about, and identify, features in the place they live and the natural world.

The coastline

- ❑ Look at pictures of the seashore that show the coastline meeting the sea: waves rolling gently onto the sand, waves crashing against cliffs and rocks and sand dunes, or show the class a video extract of the seaside. Discuss the pictures with the children. Explain that sea water wears away the cliffs and is always changing the look of the coastline. Point out how the waves leave patterns on the sand.

- ❑ Read a non-fiction book to the class about the tides (see Resources).
- ❑ Explain that the tide comes in and the waves come further up the beach; then the tide goes out and the water is further away. Talk about the things left behind on the beach when the tide goes out: seaweed, sea creatures in rock pools, shells – mussels and limpets, crabs, even small fish and shrimps. Try to show as many pictures of these as possible. (Try to show sandy beaches, rocky beaches and shingle beaches to indicate that not all beaches are of golden sand.)

The beach

Looking at water in puddles

- ❑ Take the children out into the playground on a wet, preferably windy, day to look at the puddles. If it is dry, make your own puddles by pouring a bucket of water onto the ground. Ask: Does the water stay still? Can you see how the wind makes ripples on the water?
- ❑ Inside, ask the children to blow on the surface of the water in the water tray. Can they make ripples or even small waves?
- ❑ Explain that fishermen need a safe sheltered place to bring their boats into. This is called a 'harbour'. A harbour shelters boats from the wind and rough seas. You may see fishermen tidying and cleaning their boats, cleaning their lobster pots or mending their nets in the harbour. People keep their yachts, dinghies and rowing boats safely in the harbour too.

Creative Development

Sing simple songs from memory.

Music and songs

- ❑ Sing 'One, two, three, four, five; once I caught a fish alive'.
- ❑ Sing 'I do like to be beside the seaside' (see Resources).

Role play area

- ❑ Make the backdrop for the seaside scene (see page 35).
- ❑ Make sandcastles, rock pools, crabs, starfish and a harbour (see page 36).
- ❑ Make a collection of sea shells. Look closely at the shapes and patterns. Ask the children to make observational drawings of shells.

Physical Development

Move with control and coordination.

Hoop rock pool games

- ❑ Put out some hoops on the floor. Tell the children that each hoop is a rock pool. They should pretend they are shrimps being moved along by the sea water and move around the floor avoiding the hoops. On the instruction 'Tide's out!' they must run and stand in a hoop. It does not matter how many of them are in each hoop, as long as they fit!

- ❑ As above, but a number is called out, such as 'Tide's out! Four!' Children should try to make groups of four in each hoop.
- ❑ Put out a hoop for each child. Play some sea music. The children move around the hoops, sideways as crabs. When the music stops, they race to sit in a hoop or rock pool. They must continue to move sideways when racing to a hoop. Take away one or more hoops and continue. If there is no hoop left, they are out.

Sandcastles

- ❑ Have buckets and spades in the sand tray. Add a little water so that the sand is damp. Encourage the children to make a perfect sandcastle.

Fine motor control

- ❑ Making the backdrop and details of the sea scene (see Creative Development).

WEEK 5

The role play area

The children will continue to use the role play area and will add **boats**, **flags**, **seabirds** and a **lighthouse** to the area. However, this is a science based week and all the curriculum areas will contribute to an understanding of the properties and power of water. There will be opportunities to investigate floating and sinking, surface tension, density of liquids, water pressure, pouring and dissolving.

Resources

Photocopiables:

Poems and songs (page 32)
Water experiments (pages 37–39)

Fiction book:

The Lighthouse Keeper's Lunch by Ronda and David Armitage, Scholastic Hippo (0 590551 75 2)

Non-fiction books:

Splish, Splash, Splosh! by Mick Manning, 'Wonderwise' series, Franklin Watts, (0 749656 86 7)
Water by Chris Oxlade, 'Materials' series, Heinemann Library (0 431127 34 4)
Water by Lynn Huggins-Cooper, 'Firsthand Science' series, Franklin Watts (0 749648 64 3)
Floating and Sinking by Peter Riley, 'Ways into Science' series, Franklin Watts (0 749639 57 1)

Materials:

- Pictures and posters of boats and ships
- Ship captain's hat/sailor hats
- Boat shaped sand/water tray
- Large piece of blue fabric

NB: see the instructions and materials needed for the experiments (pages 37–39).

Properties of water

Personal, Social and Emotional Development

Work as part of a group, taking turns and sharing fairly.

Circle time

- Talk to the children about being a scientist, being curious and wanting to learn and find out. Ask: What things do you like to find out about? What do you like doing at nursery or school? What do you think a scientist does? What do you like doing at home?

Discussion

- Explain to the class that they will be doing many experiments during the week and these will be set out on tables. Talk about the need to take turns and to wait patiently for their turn. Discuss the dangers of pushing or shoving in order to see what is going on; if there are too many people at a table they should move to another. Help them to produce a set of rules to remind them about taking turns, waiting for their turn, helping others, not pushing and so on.

Mathematical Development

Use everyday words to describe position.

Positional language

- Using the role play area, ask the children to place the objects in certain positions. For example, they could put the crabs in front of the rocks, behind the rocks, under the rocks and beside the rocks. Repeat with other animals or objects. Encourage them to say where they have put the object, using the correct positional language.

Extension

- Ask the children to give the instructions for placing an object in a certain position. Encourage them to think of as many different positions as possible (in front, behind, above, below, beside, under, on top, beneath).

Counting

- Count the number of blocks loaded onto foil ships (see Knowledge and Understanding of the World).

Floating and sinking

- Float card, balsa-wood and plastic (all the same size) on water. Which 'boat' can carry the most Unifix blocks/counters/paperclips?

Song

- Sing 'Five little fishes' (make up your own words).

Communication, Language and Literacy

Sustain attentive listening, responding to what they have heard by relevant comments, questions and actions.

Speaking and listening

- Share a non-fiction book about water with the class (see Resources). Ask them what they would like to know about water before showing them the book and then decide if the book has provided the answer.

Writing

- Demonstrate to the children how to complete their scientist hats (see Creative Development) by writing the sentence 'I am a scientist' around the band. Show them how to form the letters and write the words. Encourage some of them to write unaided and support others as appropriate.
- Encourage the children to make pictorial records of some of the investigations (see Knowledge and Understanding of the World) and write simple labels or captions.

Extension

- Show the group how to write up one of the experiments (see pages 37–39). Talk about the heading and help them to write a few simple sentences describing what they did and what they discovered. For example, What things float? (heading) We put a paperclip on the water. I guessed it would float and it sank. We put a plastic ball on the water. I guessed it would sink and I was wrong!

Knowledge and Understanding of the World

Investigate objects and materials by using all of their senses as appropriate.

Investigations

- ❑ Children do not always need to know the reason why something happens. They need to observe and talk and question. Make sure they know that they should not drink any of the liquids. Some activities, for example making a fountain, should only be attempted as a demonstration by an adult, as hot water is used.
- ❑ Set up all or some of the scientific experiments (see pages 37–39) and let the children move freely from one investigation to another. Adults should be on hand to ask questions, such as 'What can you see?', 'What is happening?', 'Why do you think that happened?' and 'What would happen if...?'

Scientists' hats

- ❑ Ask the children to wear their scientist hats (see Creative Development). Encourage them to take on the role of scientists. Set out some activities that need little or no supervision, such as the raisins in fizzy water activity or the target game from Week 2 (pages 13 and 16).

Flags

- ❑ Explain that many boats have flags and that children sometimes put flags on top of their sandcastles. Have a selection of ready cut, paper flag shapes: squares, rectangles and triangles with curved edges to show movement in the wind. Ask the children to design and colour the flags and fix them to lolly sticks.

Lighthouse

- ❑ Take a long, fairly wide tube and paint it in horizontal red and white stripes. Paint a door at the base in black. Add a silver, gold or clear domed top (from the junk modelling box) for the light. Fix it on the rocks in the role play area.

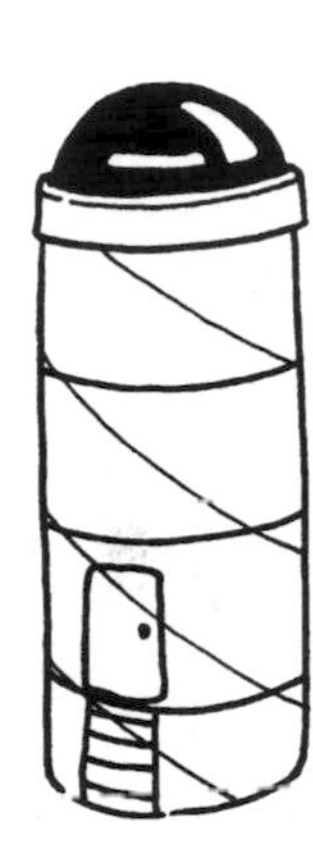

Gulls and other sea birds

- ❑ Look at pictures of sea birds. Talk about why sea birds have webbed feet. Provide a selection of paper, card and small collage materials and encourage the children to make some model sea birds. Display these in and around the role play area.

Creative Development

Use imagination in art and design.

Scientists' hats

- ❑ Give each child a strip of paper or card. Mark a line on the strip for the children to write 'I am a scientist,' (see Communication, Language and Literacy). Staple the strip to make a headband.

How to make boats

- ❑ Show the children pictures of boats: yachts, tankers, cruise liners, motor boats, rowing boats and so on. Make available a selection of junk modelling materials and encourage the children to make vessels to float on the sea. Paint and decorate them as appropriate.

Physical Development

Show awareness of space, of themselves and of others.

Action songs

- ❑ Sing and do the actions – washing the clothes to the tune 'Here we go round the mulberry bush (ring out the clothes, hang up the clothes, take down the clothes, fold the clothes, iron the clothes) ... on a cold and frosty morning.'

Shipwreck game

- ❑ Play the shipwreck game as in Week 3 (page 20).

Water

WEEK 6

The role play area

During this week the children will be able to reflect on what they have discovered and learned in the unit. Weather permitting, role play activities will be based outside. They will have beach umbrellas, a paddling pool and a boat to play in. The unit will be celebrated with a beach party at the end of the week. Children will make **flower garlands** and write **party invitations**. They will help prepare the party food, perform a poem and play games at the party.

Resources

Photocopiables:

Poems and songs (page 32)

Fiction books:

Brave Little Train by Nicola Baxter, 'Little Stories' series, Ladybird (0 721419 20 8)
You Can Swim, Jim by Kaye Umansky, Red Fox (0 099669 41 2)
Frog by Susan Cooper, Red Fox (0 099432 26 9)

Non-fiction books:

I Can Swim! by Karen Wallace, Dorling Kindersley Readers Level 1 (0 751367 96 6)
Near Water by Ruth Thomson, 'Safety First' series, Franklin Watts (0 749654 69 4)
How Should I Behave? by Mick Manning, Franklin Watts (0 749639 98 9)

Music and song:

Fantasia on British Sea Songs by Sir Henry Wood (available on many CDs)

Materials:

- Parasols and bases
- Paddling pool
- Towels (paper towels are hygienic but not as absorbent as terry towels; the children could bring in their own if necessary)
- Inflatables: lilos and rings
- A variety of fruit
- Bread and sandwich fillings
- Green jelly
- Coloured fondant icing
- Orange juice
- Sugar-free lemonade
- Sugar-free ice pops
- Fun sprinkler and hosepipe
- Volley ball, soft bat and ball, soft football, racquets and shuttlecocks
- Bucket and spade
- Bubble mixture and wire loops
- Skittles
- Skipping ropes

Beach party

Personal, Social and Emotional Development

Form good relationships with adults and peers.

Circle time

- ❑ Ask the children to describe activities they like to do on the beach. As some may not have visited the seaside, have some pictures of beach scenes to show them.
- ❑ Talk about holidays. Where are the children going for their holiday? Will they be going away or staying at home? Will they be going swimming in the sea or the swimming pool?
- ❑ Talk about being safe in water and doing as they are told. Make them aware of the power and danger of the sea.

Discussion

- ❑ Explain that they are going to have a beach party and are going to invite other people to come to it. Talk about manners and making people welcome. Practise saying, 'Welcome to our beach party – please come this way.' Remind them about saying 'Please' and 'Thank you' when they are offered food.
- ❑ Share a book about manners (see Resources).

Mathematical Development

Begin to relate addition to combining two groups of objects and subtraction to taking away.

Additing and subtracting

- ❑ Using the skittles with numbers on the bases (see Physical Development), ask the children to add up the score of the skittles they have knocked over.
- ❑ Add two or more objects; for example, the fish in the role play area. How many are left if you take two away?

Extension

- ❑ Ask the children to add or subtract two or more different objects. Help them to display these as sums; for example, $2 + 3 = 5$ and $4 - 2 = 2$.

Colour sorting

- ❑ Ask the children to sort the party fruits by colour.
- ❑ Ask them to find examples of different blues or greens in the role play area.
- ❑ Using a pegboard, ask the children to place different colours and numbers in rows; for example, 'Can you put four red pegs in the top line?'

Counting

- ❑ Many of the games (see Physical Development) involve the children counting. Encourage them to help each other to count accurately.

Measuring

- ❑ Show the children how to measure accurately. Talk about measuring the water for making the jelly (see Knowledge and Understanding of the World) and what would happen if you used too much or too little.
- ❑ Ask the children to measure different amounts into containers in the water play tray.

Communication, Language and Literacy

Speak clearly and audibly with confidence and control and show awareness of the listener.

Listening

- ❑ Read *You Can Swim, Jim* or another story about learning to swim (see Resources).

Speaking

- ❑ Tell the children that they are going to recite a poem and sing a song to the guests. This could be the poem they have written in class or a favourite one from page 32; for example, 'There are big waves'. Help them to decide which poem and song they would like to do and practise this as necessary.

Writing invitations

- ❑ Explain to the class that they are going to have a beach party and they can invite other people or children to join them. Talk about writing invitations and what information is needed on the invitation.
- ❑ Draw an outline of a palm tree on the board and write an invitation on it, talking about the writing process as you write.

Beach party

- ❑ Draw an outline of a palm tree on A4 paper and photocopy it. Ask the children to cut out the tree and glue on a few coconuts. Then ask them to write the invitation. Support their writing as appropriate, for example by using highlighter pen for them to go over in pencil.

Extension

- ❑ Ask some of the children to label the fruit that they will eat. Ask them to identify the first letter sounds of the different fruits and show them how to write the words.

Knowledge and Understanding of the World

Select the tools and techniques they need to shape, assemble and join the materials they are using.

Select a variety of fruits for the party (see Creative Development). Choose familiar fruits as well as more exotic and less familiar ones for the children to try. Help the children to prepare the fruit for the party and place it on plates. Label the plates (see Communication, Language and Literacy).

Swimming and safety

- ❑ Emphasise the need for care in or near water, as discussed in Personal, Social and Emotional Development. Talk about being careful not just by the sea but also around ponds and lakes, particularly on holiday.

Party food and drink

- ❑ Make 'Undersea jelly'. You will need a green jelly, coloured fondant icing and a clear bowl. Make the jelly, following instructions on the packet. Mould sea creatures (fish, octopus, eel, starfish and so on) out of fondant icing. Stick some creatures around the edge of the bowl with a tiny amount of water. When the jelly has set, chop it up with a fork and spoon into the bowl, taking care not to disturb the sea creatures. Add more creatures in the middle of the jelly and place a couple more on the top.
- ❑ Make sandwiches. Encourage the children to choose the fillings beforehand and make sandwiches on the morning of the party. If you have any cutters shaped like fish or other sea creatures, use them to cut fun shaped sandwiches.
- ❑ Make a cocktail drink. You will need a carton of orange juice, low sugar lemonade, slices of orange and ice cubes. Mix together just before the party begins.

Music and song

- ❑ Sing 'Jelly on the plate'.

Creative Development

Respond in a variety of ways to what they see, hear, small, touch and feel.

Beach party

- ❑ Tell the children that they are going to have a beach party and that on the day they will be able to come to school in shorts, T-shirts and 'shades'. (You could link this to 'Shades Day' fund raising event for Guide Dogs for the Blind – www.guidedogs.org.uk) Explain that their party will take place on a tropical island where it is warm and sunny. On this island, lots of beautiful flowers grow, so the islanders like to make necklaces of flowers to wear.

Making flower garlands

- ❑ Give each child a plastic needle, threaded with a length of wool. Have ready cut circles of tissue paper. Show them how to take two colours of tissue and thread onto the wool. Lightly scrunch each 'flower'. Tie the ends to make the garland. The children may wish to make a flower to wear in their hair!

Party invitations

- ❑ Create party invitations (see Communication, Language and Literacy).

Observational drawings

- ❑ Provide some fruits for the children to study and then make observational pencil drawings of. Talk about shape and colour. When the sketches are finished, tell the children to paint over the pencil with a wash of the appropriate colour.

Outside role play area

- Paddling pool and sun area – set up a paddling pool, in the shade if possible, and place chairs around it. Fill the pool and add a little baby bath liquid. Remind the children to take off their shoes and socks. Tell them to sit on chairs and dip their feet into the water. Can they make more bubbles by splashing their feet? What does the water feel like? Add small objects, such as balls, corks and plastic plates. Can they pick up these objects with their feet?

  NB: an adult must supervise this activity at all times and rules should be explained carefully beforehand. For example, you do not stand up in the water, you should stay sitting on your chair, do not splash so hard that you make others wet and wait until the teacher or another adult says you can go and dry your feet.

  Place parasols around the boat and the paddling pool. If possible, place more beach umbrellas in the playground. Under one umbrella have a box of books related to the unit. Inflate lilos and rings and place them as seats in the play area. Bring all the dressing-up clothes and props from the role play area outside. As a real treat, give each child a sugar-free ice pop at the end of the party – a prize for all who took part in the party games!

Physical Development

Move with control and coordination.

Playing beach games

- Volley ball. Draw a chalk line on the playground to represent the net. Ask the children to try to pat a large soft ball over the net to their partner. Variations: bounce the ball over the line; throw and catch; roll the ball.
- Cricket. A game for about four to six children. You will need a small soft ball and a bat. Mark stumps with chalk on a wall and place a cone or marker a short distance from the stumps. Mark a circle for the bowler. Bowler rolls the ball to try to hit the stumps. Batsman tries to hit the ball and run to the cone and back (this counts as one run). Fielder who retrieves the ball is the new bowler. Batsman is out if the ball touches the stumps. Fielders can run with the ball or throw to one another. The ball can only be aimed at the stumps from the bowler's circle.
- Bat and ball. Leave out a box of bats and balls for the children to use as they please.
- Beach football. If you have the space, designate an area for football. Mark two goals. Provide a soft football and allow the children to organise their own game. You may need to establish rules. For example, the ball should be kicked along the ground and not high in the air. You may also need to restrict the number of players for safety. Try to encourage girls to take part!
- Sandcastles. Provide buckets and spades for the sand tray or sandpit. Encourage the children to work together to build sandcastles. Can they make them as elaborate as possible? A prize for the best? Take digital photos of efforts – the children could take them.
- Bubbles. During the party, let the children play with bubbles.
- Skittles. Set up a skittles set or make your own from empty drinks bottles. Write a number on the base of each bottle.
- Badminton. Try to get hold of an inexpensive, children's badminton set (Woolworth and most toyshops) or some old racquets and shuttlecocks. If possible, string a net between poles or trees. No need to set rules as the children will be concentrating on making contact with the shuttlecock!
- Long jump. Mark a chalk line on the ground for the start. Children to start with their feet behind the line and jump, both feet together, as far as they can. Mark the jumps with a coloured X. Who can jump furthest?
- Hurdles. Set up a course of low hurdles with skipping ropes laid over, for example, cones. Make sure that if the rope is touched it will fall easily and not trip the runners. This is for fun, not a race.
- Running through the spray. Use a hose and sprinkler attachment and set it up in the playground for the children to 'run through'. Ideally, use an attachment which can be staked in the ground and which has a flexible hose with a novelty 'head' such as a daisy. The head spins and jumps at random, while spraying a fine jet of water.

  NB: this activity needs careful supervision and is safer if the sprinkler is sunk into grass but sprays onto a hard surface, as wet grass becomes slippery very quickly!
- Hornpipe. Show children how to dance a hornpipe! – a sailors' dance of old – dancing with arms folded in front! Use suitable music, such as Sir Henry Wood's 'Fantasia on British Sea Songs' (see Resources).

Review and evaluation

Ask the children to reflect on the topic. What have they enjoyed learning about? Which was the most exciting experiment? What did they like making the most? Which games did they like playing?

Poems and songs

Seaside

Sand in the sandwiches,
Sand in the tea,
Flat, wet sand running
Down to the sea.
Pools full of seaweed,
Shells and stones,
Damp bathing suits
And ice-cream cones.
Waves pouring in
To a sand-castle moat.
Mend the defences!
Now we're afloat!
Water's for splashing,
Sand is for play,
A day by the sea
Is the best kind of day.

Shirley Hughes

The seaside

Are we nearly there?
Can you see the sea?
Who will be ready first?
Me! Me! Me!

Does the sand tickle
Down by the sea?
Who can make footprints?
Me! Me! Me!

The seagulls are crying,
'Shush' says the sea.
Who dares put a toe in?
Me! Me! Me!

Jo Peters

I hear thunder

I hear thunder, I hear thunder
Hark don't you? Hark don't you?
Pitter-patter raindrops, pitter-patter raindrops
I'm wet through, so are you.

Anon

Water in bottles, water in pans

Water in bottles, water in pans
Water in kettles, water in cans;
It's always the shape of whatever it's in
Bucket or kettle, or bottle or tin.

Rodney Bennett and Clive Sansom

If you ever

If you ever, ever, ever, ever, ever
If you ever, ever, ever meet a whale
You must never, never, never, never, never
You must never, never, never touch its tail.
For if you ever, ever, ever, ever, ever
If you ever, ever, ever touch its tail
You will never, never, never, never, never
You will never, never meet another whale.

Anon

The friendly octopus

Eight arms for me, eight arms for me,
I'm a friendly octopus, under the sea.

I've got
One arm to blow my nose,
One arm to wave with,
One arm to brush my teeth,
One arm to shave with,
One arm to comb my hair,
One arm to shake with,
One arm to blow a kiss,
One arm to eat a cake with.

Eight arms for me, eight arms for me,
I'm a friendly octopus, under the sea.

Mike Jubb
(www.mikejubb.co.uk)

How to make the background for the role play area

You will need: large sheet(s) of backing paper, either white or blue, paints, large brushes (for example, decorating brushes), newspaper or plastic sheeting to cover the floor or ground, coloured and plain Cellophane, crepe paper, tissue paper, bubble wrap, metallic felt-tipped pens, green garden netting.

This background is created using wet on wet technique and is most effective if the backing paper can be taped or pegged to a wall or fence outside, so that the paint can drip downwards. You will need to protect the surrounding wall and ground with newspaper or plastic sheeting and ensure the children wear aprons! If it is not possible to work outside, paint on a table and try lifting the backing paper occasionally to allow the paint to drip down. Don't forget to protect the floor!

Thoroughly dampen the paper with large brushes and water. While the paper is wet, paint shades of blue and green, allowing the colours to spread, merge and drip downwards. Allow to dry. Roughly cut strips of clear, blue and green Cellophane and stick, vertically, to the background in random areas. Cut circles out of bubble wrap, various sizes, and stick in wavy vertical patterns, largest at the bottom, gradually getting smaller. Mount the backing paper to line the walls of the role play area.

How to make the underside of a boat

Staple together large sheets of brown paper, then fold in half and draw a boat shape. Cut out and staple the sides. Fold the base of the boat inwards. Glue brown or wooden beads to the underside to represent barnacles. Suspend the boat from the ceiling.

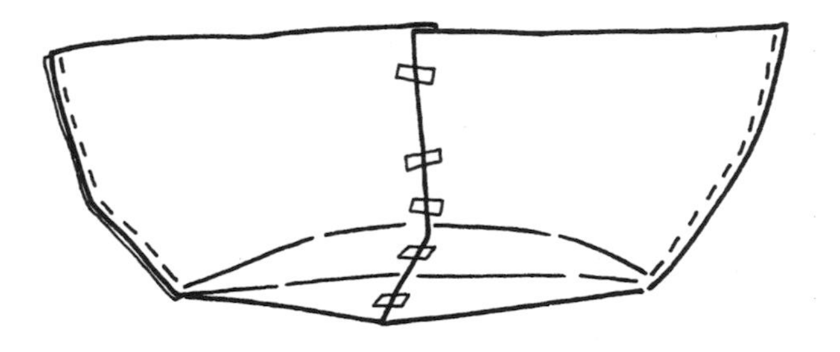

How to make fishing nets

Drape and suspend garden netting from the ceiling and around the boat. Leave the ends of the netting to drape over the background.

How to make rocks on the sea bed

Cut large rock shapes from card. Stick scrunched kitchen paper over them and paint with shades of grey paint mixed with a little PVA glue. Stick to the background at floor level on one side of the role play area.

How to make seaweed

Roughly cut long strips of brown, red, or green Cellophane. Cut circles of the same and glue randomly to the strips. Suspend the seaweed from the ceiling.

How to make coral reef

Cut small rock shapes out of coloured card. Decorate using some of the following ideas: cover with dots, using metallic pens; cover with art straws to give a spiky appearance and paint with bright colours; cover with brightly coloured, scrunched tissue paper; cover with scraps of coloured Cellophane and tissue paper; cover with scraps of crepe paper; cover with flower shapes (circles of tissue paper, folded into four and the edge snipped). Stick to background at floor level on one side of the role play area.

How to make handprint fish

Make paint handprints on scraps of coloured card or paper. Cut out and turn sideways. Add a google eye and a mouth.

Making things 2

How to make 3-D fish

Put a sheet of laminating film through the laminator. Cut into strips. Staple fish to one end and staple the top of the strip to the background. This will give a 3-D effect of fish swimming in the water.

Cut double fish shapes out of strong paper. Staple round the edges, leaving enough room to stuff with newspaper. Staple the remaining edge. Decorate with scraps of shiny paper, Cellophane and tissue. Add eyes and a tissue paper streamer tail. Suspend on fine thread from the ceiling.

How to make Cellophane fish

Cut out, from coloured Cellophane, some fish shapes. Decorate with metallic markers. Display in shoals, hanging from the ceiling.

How to make CD fish

Take an old CD and add a heart shaped mouth cut from coloured paper. Add fins and fan tails.

How to make an octopus

Draw an outline of an octopus and cut small circles of coloured paper. Tell the children to cut out the octopus and stick the circles along the tentacles to represent the octopus suckers.

How to make eels

Stuff long socks with newspaper and secure with an elastic band. Cut and glue felt eyes and a mouth.

How to make jelly fish

Roughly cut semicircles of bubble wrap. Glue thin strips of bubble wrap to the straight side. Fix to the role play background near the surface.

How to make small fish

You will need some ready cut scales, dull blue, green and grey and some ready cut shiny scales. Ask the children to draw fish on blue or green card and cut them out. Decorate with coloured scales. Add one shiny scale.

How to make an octopus costume

You will need a wicker lampshade with the fittings removed or a round basket (the children need to be able to see through the wickerwork), old brown tights, material for stuffing (lightweight fabric scraps, foam, polyester, even newspaper will do), felt scraps and green or brown strips of crepe paper.

1. Cut eight legs/tentacles from the tights and lightly stuff.
2. Cut out two felt eyes and a felt mouth and stick towards the top of the lampshade.
3. Staple the tentacles firmly around the lampshade. Staple from the inside of the lampshade to avoid the children being scratched when they wear the costume.
4. Attach strips of green and brown crepe paper to the top of the octopus to represent seaweed.

How to make a rainbow fish costume (by an adult)

Take a large circle of fabric, any dull colour, about one metre in diameter, and cut a slit in the centre so that it fits over a child's head. Stick self-adhesive Velcro patches (hook) randomly around the 'poncho'. Cut scale shapes out of shiny paper and laminate. Stick a Velcro patch (loop) to the back of each scale. Fix the scales to the 'poncho'. These can be given away when the children are re-enacting the story of the Rainbow Fish.

How to make badges for the children to wear as little fish

Put a sheet of laminating film through the laminator and cut out scale shapes. Tape a safety pin firmly to the back and stick a Velcro patch (hook) on the front of each one. The little fish can fix their new shiny scales onto their badges.

How to make fish hats

Cut two semicircles out of thin A3 card or firm paper. Staple (adult to do) around the curved edge to join, leaving the base open to fit over the head (adjust as necessary). Cut a small circle (15cm in diameter) out of paper and make a slit in the middle. Fold in half and stick to the fish for the mouth (you will need to fold corners in a little). Make fins and tail out of tissue or crepe paper. Add pompom eyes. Encourage the children to wear the hats during imaginative play. Can they move together as if in a shoal?

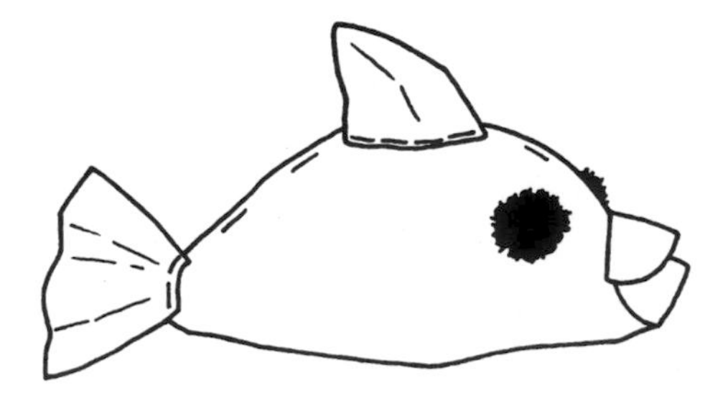

How to make portholes: looking through a porthole under the sea

You will need two paper plates for each child, a selection of collage materials, some fine thread and cellophane, clear or coloured.

1. Cut the middle out of one of the plates. Paint the outsides of both plates grey. (Don't let the children make the plates soggy!) Cut a circle of Cellophane and glue to the inside (the eating surface) to cover the hole. On the inside of the other plate, create an underwater scene. For example, paint the plate blue, add tissue seaweed, grey paper rocks, small shells and so on. Cut a sea creature shape out of thin card (this could be an octopus, shark, dolphin or rainbow fish). Suspend it on fine thread from the top of the plate over the scene.
2. Staple the two plates together (eating sides together).
3. Draw or paint black dots round the front, to represent rivets.
4. Suspend the portholes around the classroom.
5. Suspend a few in, or at the entrance to, the role play area.

How to make the background for the seaside

You will need one long sheet of deep blue backing paper such as 'milskin', enough to line the wall of the role play area.

1. Mask off a narrow strip at the top, with either masking tape or newspaper, to keep a crisp horizon line.
2. The children paint with large brushes. Paint sweeping horizontal strokes, starting at the top with dark blue paint. Continue underneath with a lighter blue and finally pale blue paint at the bottom.
3. When dry, give the children oil pastels and wax crayons (shades of blue). Ask them to look for 'waves' along the brush strokes and to draw a series of dots along the curves, dark at the top, getting lighter towards the bottom of the background.

Making things 4

4. Mix some gold or bronze metallic paint with a little blue in a shallow tray (keep the paint quite thick). Tell the children to step, barefoot, in the tray and walk across the top of the background, heel to toe. Repeat over the dark blue area. Support each child carefully throughout this process as the paint will be very slippery.
5. Under the rows of footprints, make rows of handprints. Mix some gold or bronze metallic paint with a different shade of blue, perhaps a lighter turquoise blue. The children print with their hands to make wavy lines, by placing hands first in one direction, then the next row in the other. The first line should merge into the footprints.
6. On the pale blue at the bottom of the background, make fingerprints. Mix a little metallic paint with a very pale blue. The children dip fingers of one hand into the paint and make random patterns.
7. Take black pastels and scribble small areas in between the printing to give the appearance of depth in the sea.
8. Highlight small, lighter areas with white chalk.
9. Dip a toothbrush into some white paint and by drawing a ruler over the brush, splatter paint at the bottom to represent spray from the waves.
10. Glue a few sequins to the sea.
11. Carefully remove the masking tape or newspaper at the top. Draw a fine line along the horizon line with white chalk and smudge with finger.
12. Line the wall in the role play area.

How to make sandcastles

Help the children to build sandcastles out of boxes and cylinders from the junk modelling box. Cover the sandcastles by sticking strips of newsprint or other thin white paper over the top. Paint with sand coloured paint. Paint a thin layer of PVA glue and sprinkle sand over the surface. Place a sheet of yellow or beige fabric over the floor in the role play area and place the sandcastles on top.

How to make rock pools

Tear or cut rock shapes from foam scraps and paint them grey. Stick or staple them to the background at floor level. Place pieces of silver foil or bubble wrap in between to represent pools. Stick a few shells on the rocks. Splatter some more white paint over the top of the rocks.

How to make crabs

Show the children pictures of crabs. Look at the shell and pincers. You will need corks, pipe cleaners, beads and scraps of card.

1. Tape pipe cleaner legs along the length of a cork (five pairs). Cut two pincer shapes and stick to front legs. Glue two bead eyes to the front. Cut a heart shape from card and stick on top of cork.
2. Place crabs in and around the rock pools.

How to make starfish

Show pictures of starfish and ask the children to draw them on orange card and cut them out.
Show them how to dribble lines of PVA glue, radiating from the centre like the patterns on a starfish. Sprinkle sand over the glue and allow to dry. Place some in the rock pools in the role play area and display the others around the classroom.

How to make sea defence or breakwater

Explain the purpose of a breakwater.
Help a child to cut a long, narrow triangle of brown paper or card. Glue to the base of the background.

How to make a harbour

Paint three or four boxes grey and fix to the background at floor level. On top of one of these boxes drape some netting and place a shallow tray containing fish on top. To make the fish, mould kitchen foil into fish shapes and lay in rows in the tray.

Experiments to find out about the surface tension of water

Experiment 1

- You will need: a clear container/cup of water, 1p coins

1. Fill the cup to the very top. Carefully slide coins into the water, one at a time.
2. Watch what happens to the surface of the water. (The water should bulge as if it has a kind of skin.)
3. What happens if you put in more coins?

Experiment 2

- You will need: water, a clean surface, drinking straw, spoon, magnifying glass, washing-up liquid

1. Dip the spoon into the water and drip onto the surface.
2. Look at the droplets with a magnifying glass. What shape are they?
3. Dip the straw into some washing-up liquid and touch the droplets. What happens?

Experiment 3

- You will need: a shallow dish of water, paperclips, kitchen foil, dropper or straw

1. Make some pond skaters. Cut two small squares of foil. Place a paperclip in the centre of one and cover with the other. Make two snips on either side and carefully mould the foil to shape legs.
2. Tell the children to very carefully put the pond skaters onto the water. To make this easier, put the pond skater on a piece of tissue paper (single layer of facial tissue) and, holding the tissue firmly, place it gently on the surface of the water. The tissue will sink, leaving the insect afloat.

3. Drip washing-up liquid onto the water and observe the skaters dart away. (If you have a pond or container sunk into the ground, supervise children observing water boatmen or other insects walking on the water's surface.)

Experiment 4

- You will need: a shallow container of water, some pepper, washing-up liquid

1. Sprinkle pepper on the water.
2. Drip washing-up liquid on the surface and watch what happens.

Experiment 5: Blowing bubbles

- You will need: a container with a solution of one part water to three parts washing-up liquid, loops made out of wire (florists' wire or other thin wire) – circle, square, diamond shapes and so on

1. Dip loops into soapy mixture and blow gently. What happens?
2. Try different shaped loops. Do they make different shaped bubbles?
3. Try to catch the bubbles. What happens?

Experiments to find out about water density

Experiment 1: Mixing liquids

- You will need: a clear drinks bottle containing a little cooking oil and mostly water

1. If possible, let the children see you putting the liquids into the bottle.
2. Shake the bottle and allow the liquids to settle.
3. What happens? Why do you think this happens?
4. Relate this to oil spillages on the sea. Oil that is washed ashore makes beaches very dirty and harms wildlife. Ask: How do you think the oil could be cleaned up? Relate this to washing up the dishes at home.

Experiments 2

Experiment 2: Drip test

(This experiment works best if two children work together.)

- You will need: a large safety mirror, water, cooking oil

1. Each child dips a brush into one of the liquids.
2. Say, 'Ready, steady, GO!' On 'GO', drop liquid onto the top of the mirror.
3. Incline the mirror slightly and watch what happens. Which liquid runs down the mirror faster/slower?

Experiment to find out which materials dry faster than others

Washing line

- You will need: pieces of fabric – cotton, wool, velvet, plastic, silk etc, bucket of water, washing line, pegs

  Tell the children to soak materials in the water and hang them on the line to dry. Which materials dry the quickest? Ask what would happen if they hung out the clothes on a dull day. Would they dry as fast as on a sunny day? Why is there a difference? (Remind them about evaporation in Week 1.)

Experiments to show water pressure

Experiment 1

- You will need plastic drinks bottles with three holes pierced through one side at varying heights, a bucket of water

1. Dip a bottle into the water to fill.
2. Hold the bottle over the bucket. What happens?
3. Why do you think the water spurts a longer jet through the hole at the bottom?

Experiment 2: Make a fountain

- You will need: a small plastic screw-top bottle (make a hole in the lid, large enough to allow a straw to be pushed through), Plasticene or playdough to seal the hole around the straw, a large bowl or container, food colouring

1. Half fill the bottle with water and a few drops of food colouring.
2. Firmly screw on the lid with the straw in place.
3. The end of the straw should be well underneath the coloured water level.
4. Stand the bottle in the large container and almost fill with hot water.
5. Stand back and watch what happens.

(The hot water warms the air in the bottle. As the air expands, it pushes the water in the bottle up through the straw to make a fountain spray.)

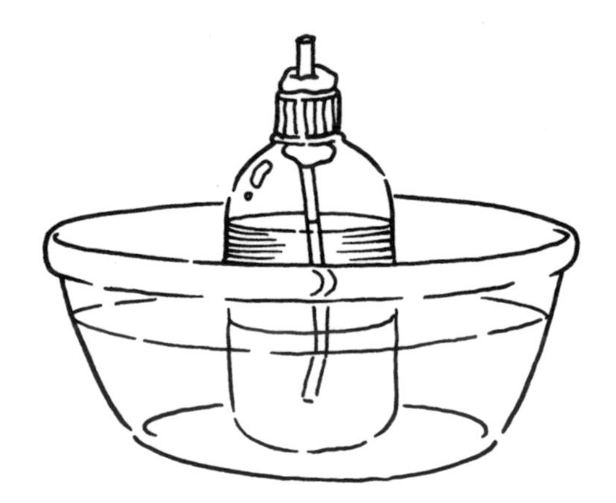

General experiments

Experiment 1: Can water flow uphill?

- You will need: cups, a bucket of water, a slope or ramp

1. Ask the children to pour water at the top of the slope and observe the flow.
2. If they pour water at the bottom of the slope, will the water flow up?
3. Explain canals and locks in simple terms. If possible, set up Aquaplay or a water tray with a lock gate to illustrate the principle of locks.

Experiment 2: Water tray

- You will need a full water tray, a selection of scoops, spoons, containers, sieves, funnels, waterwheels, paddles and so on

1. Let the children play and experiment freely.

Experiment 3: What substances dissolve in water?

- You will need: clear lidded containers, white sugar, brown sugar, salt, butter, rice

1. Ask the children to predict which of the ingredients will dissolve in water.
2. Add one ingredient to each container of water. Screw the lids on tightly. Shake the containers. Were they right?

Experiment 4: Syphoning water

(Transfer water from one container to another without pouring!)

- You will need: a length of plastic tubing, two containers, water

1. Place a full container of water on a table and an empty container on the floor.
2. Place one end of the tube in the container of water.
3. Put your finger over the other end and let go over the empty container. What happens?
4. Vary the height of the containers. Does this make a difference?

Experiments to see which objects float and which sink

Experiment 1: Different materials

- You will need: a water tray, a selection of objects – stone, plastic ball, twig, ruler, paperclip, coins, nail, lolly stick, matchstick, foil and foil dishes.

1. Ask the children to find out which objects float and which sink.
2. What happens if you screw up a ball of foil?
3. What happens if you make a dish out of foil?
4. Can they identify which materials float/sink?

Experiment 2: Boats

- You will need a water tray, foil dishes, plastic blocks such as Unifix

1. The children should investigate how many blocks can be loaded onto foil dish ships before they sink.
2. Explain that real ships have a mark along the side called a Plimsoll line, which shows the limit for loading passengers or cargo.

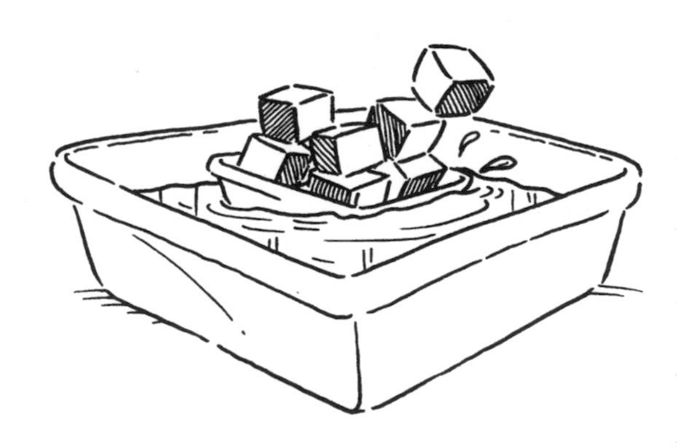

Observational Assessment Chart

Unit: ____________ Class: ____________ Date: ____________

Name	Personal, Social and Emotional Development	Communication, Language and Literacy	Knowledge & Understanding of the World	Mathematical Development	Creative Development	Physical Development
	Y B G ELG	Y B G ELG	Y B G ELG	Y B G ELG	Y B G ELG	Y B G ELG
	Y B G ELG	Y B G ELG	Y B G ELG	Y B G ELG	Y B G ELG	Y B G ELG
	Y B G ELG	Y B G ELG	Y B G ELG	Y B G ELG	Y B G ELG	Y B G ELG
	Y B G ELG	Y B G ELG	Y B G ELG	Y B G ELG	Y B G ELG	Y B G ELG
	Y B G ELG	Y B G ELG	Y B G ELG	Y B G ELG	Y B G ELG	Y B G ELG
	Y B G ELG	Y B G ELG	Y B G ELG	Y B G ELG	Y B G ELG	Y B G ELG
	Y B G ELG	Y B G ELG	Y B G ELG	Y B G ELG	Y B G ELG	Y B G ELG
	Y B G ELG	Y B G ELG	Y B G ELG	Y B G ELG	Y B G ELG	Y B G ELG
	Y B G ELG	Y B G ELG	Y B G ELG	Y B G ELG	Y B G ELG	Y B G ELG
	Y B G ELG	Y B G ELG	Y B G ELG	Y B G ELG	Y B G ELG	Y B G ELG

Circle the relevant Stepping Stones (Y = Yellow; B = Blue; G = Green or ELG = Early Learning Goal) and write a positive comment as evidence of achievement.